ON THE FARM

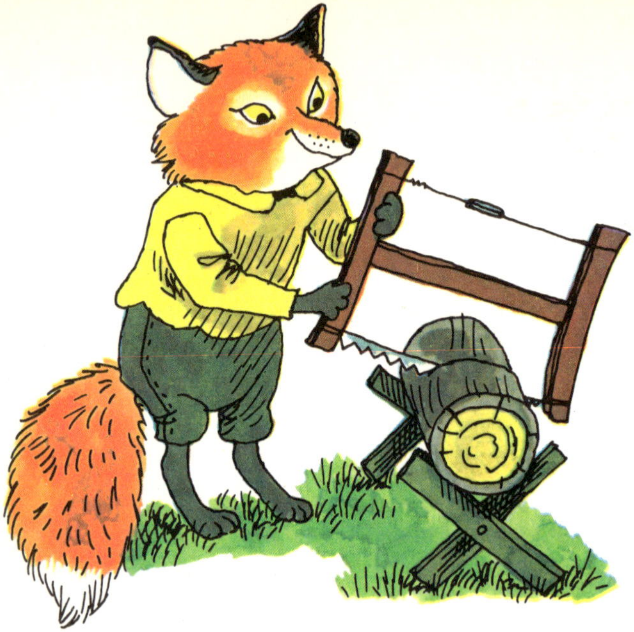

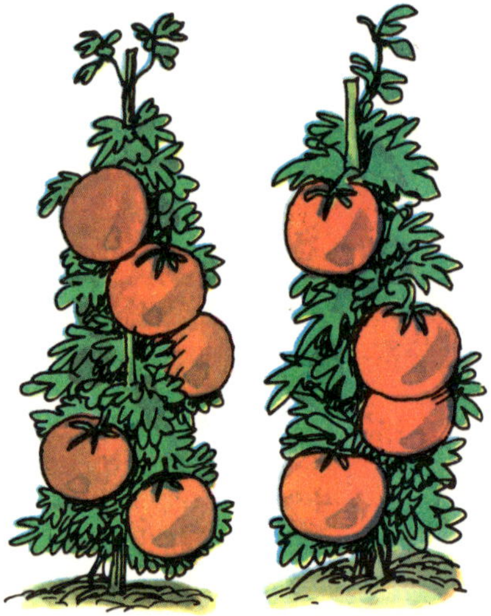

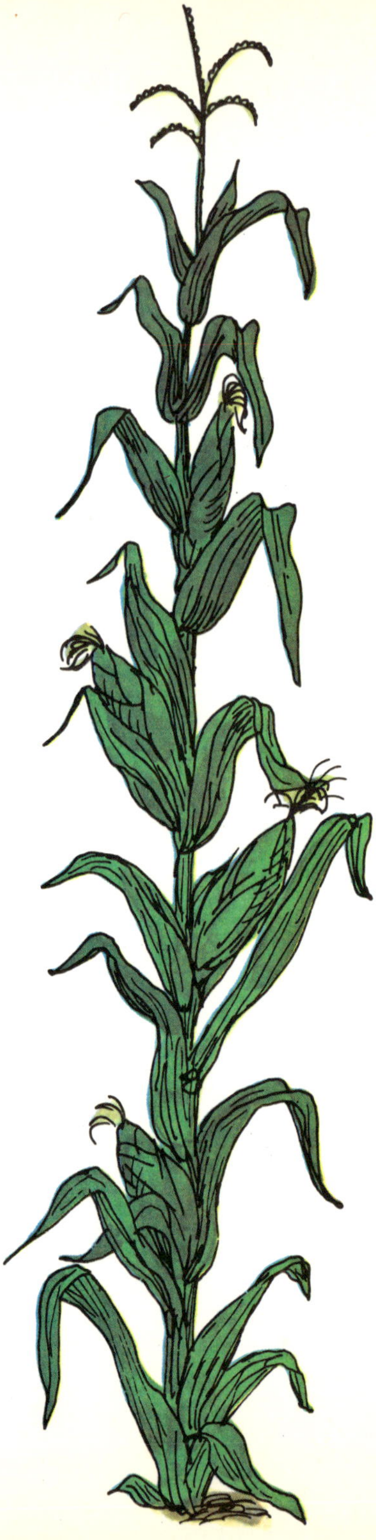

ON THE FARM

with illustrations by
Richard Scarry

Golden Press • New York
Western Publishing Company, Inc.
Racine, Wisconsin

Copyright © 1976, 1963 Western Publishing Company, Inc. All rights reserved.
No part of this book may be reproduced or copied in any form without written permission from the publisher.
Printed in the U.S.A.
GOLDEN, A GOLDEN BOOK® and GOLDEN PRESS® are trademarks of
Western Publishing Company, Inc.
Library of Congress Catalog Card Number: 75-27068

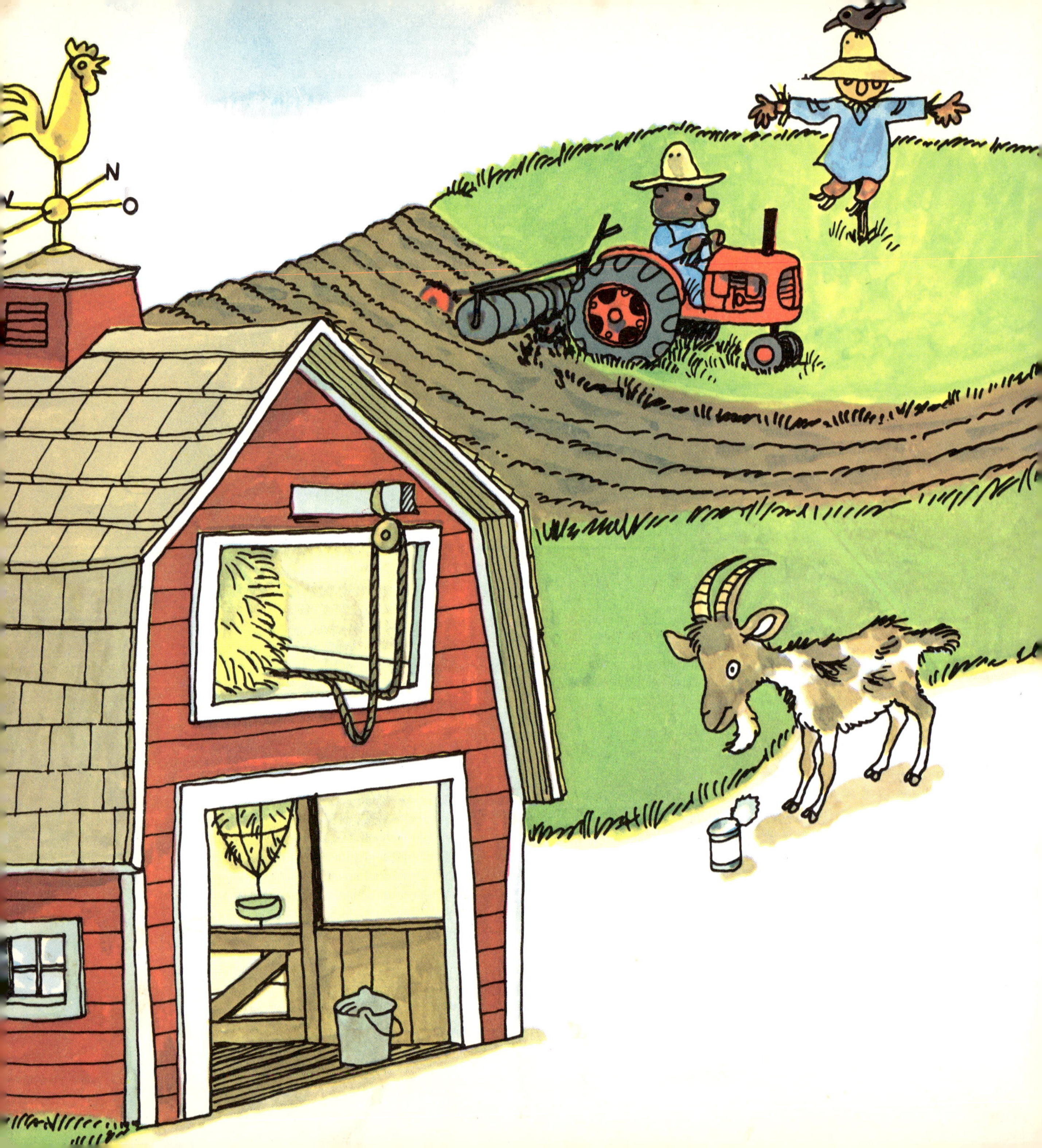
N
O

Farmer Bear

Farmer Bear is a farmer. This is his farm. All year round, he works on the land. Can you see him riding on his tractor? He is pulling a disk harrow. It turns up and mixes the soil.

Farmer Bear has a big red barn to keep his animals and machinery in. Do you see some hay in the hayloft? That is what Farmer Bear feeds his cows and horses when they are not out at pasture.

Do you see:

pickup truck	milk can
weathervane	tires
scarecrow	barn
Farmer Bear	pail
tractor	crow
goat	hay

Wind Power

Do you know how a windmill works? The wind turns the sails and the sails turn the machinery inside. Farmer Bear uses the turning machinery to grind grain or run other machinery.

Point to:

hen
duck
well

bee
pitchfork
ducklings

beehives
pond
windmill
chicken coop

The Duck Pond

Fish live in the pond, but you can't see them. They are under the water. You can see the ducks, though. They like to swim in the pond.

Farmer Bear gets eggs from his ducks and chickens. He gets honey from his bees and fresh water from the well.

WATCH OUT FOR BEES!

Work on the Farm

Do you see:

Rabbit	Owl
hoe	hose
seeds	spade
spreader	Pig
crow	rake
Raccoon	garden fork

Farmer Bear's friends like to help on the farm. Today they are planting a garden. Rabbit is raking away the pebbles. Pig is hoeing a furrow. Raccoon is planting the seeds. I wonder what vegetable they are planting?

Owl wants to help, too. He wants to water the new garden, but he has forgotten something. Can you guess what? He has forgotten to turn on the water!

Plants

All sorts of plants grow on Farmer Bear's land. Farmer Bear plants some of them, and some of them grow wild.

Bees collect nectar from the flowers. They use it to make their honey.

Other insects eat the plants and stop them from growing. Rabbit is spraying the plants to protect them from that kind of insect.

Do you know the names of the plants in the picture?

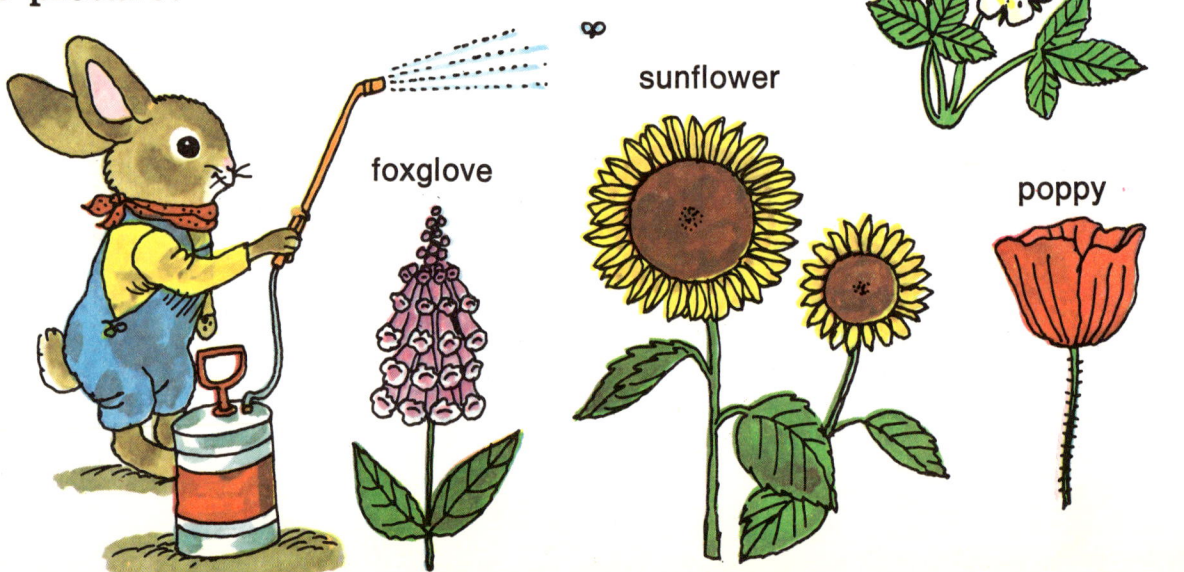

buttercup

clover

strawberry

foxglove

sunflower

poppy

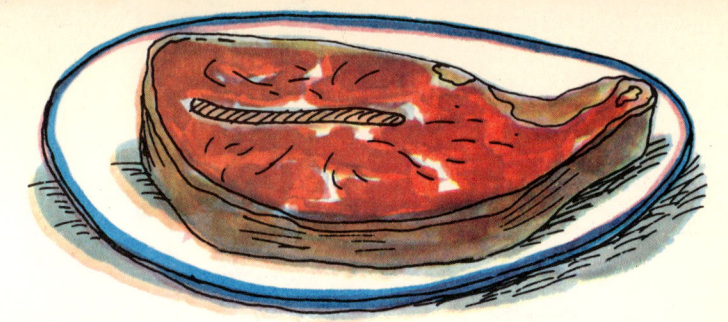

Hearty Appetite!

All the farm work makes Farmer Bear and his friends very hungry. They eat a big dinner every day. Here are some of the foods they like to eat.

Did you know that all of these foods came from the farm?

What do you like to eat when you are very, very hungry?

What Weather!

A storm is coming. The workers hurry in from the fields. Flash! goes the lightning. Crash! goes the thunder. Down comes the rain.

Soon the storm is over.
The workers go back to work.
The wind is still blowing and the children are flying their kite.
Oh dear! Little Cat is flying his hat, too. Hurry, Little Cat, before your hat blows into the duck pond!

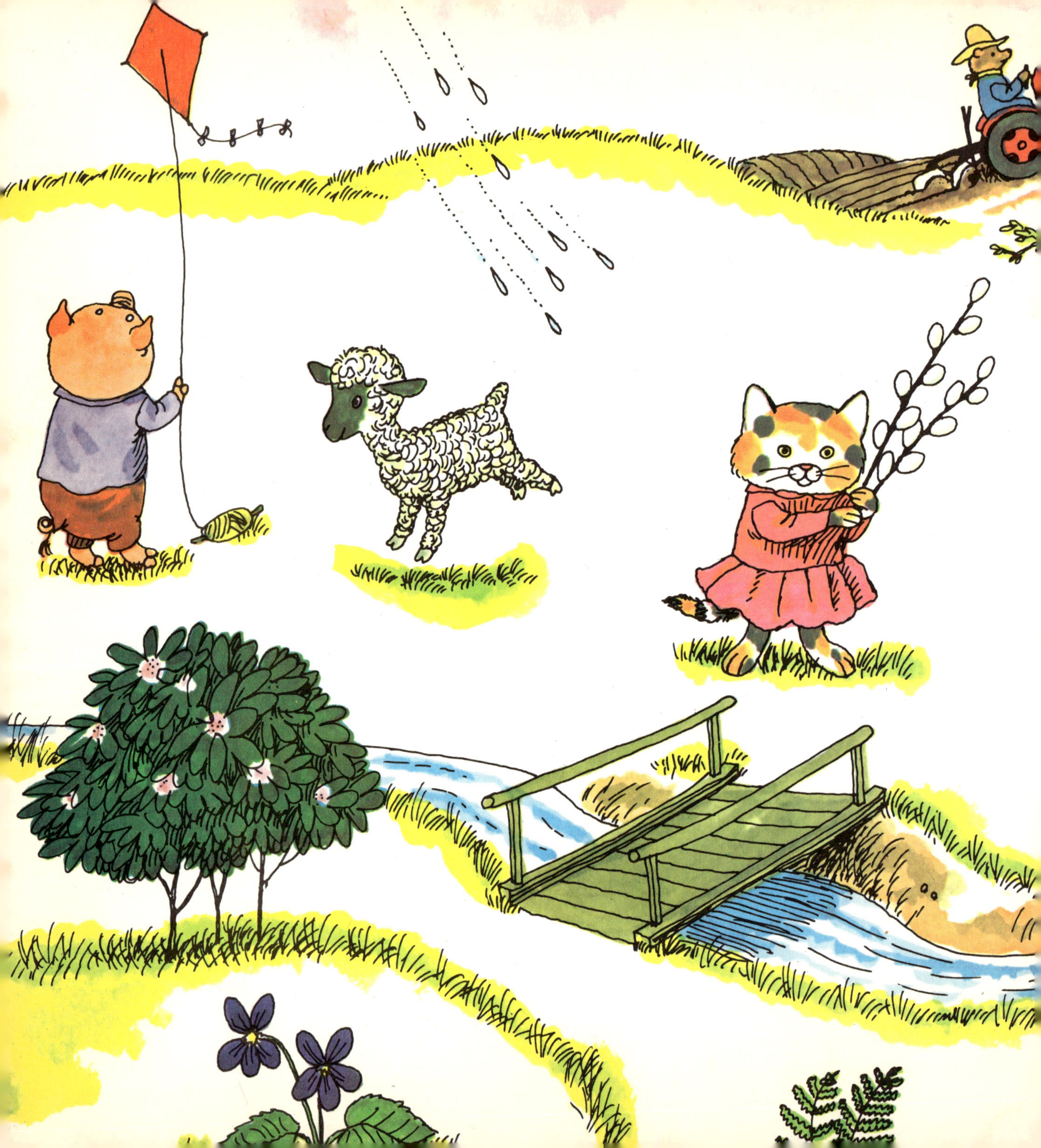

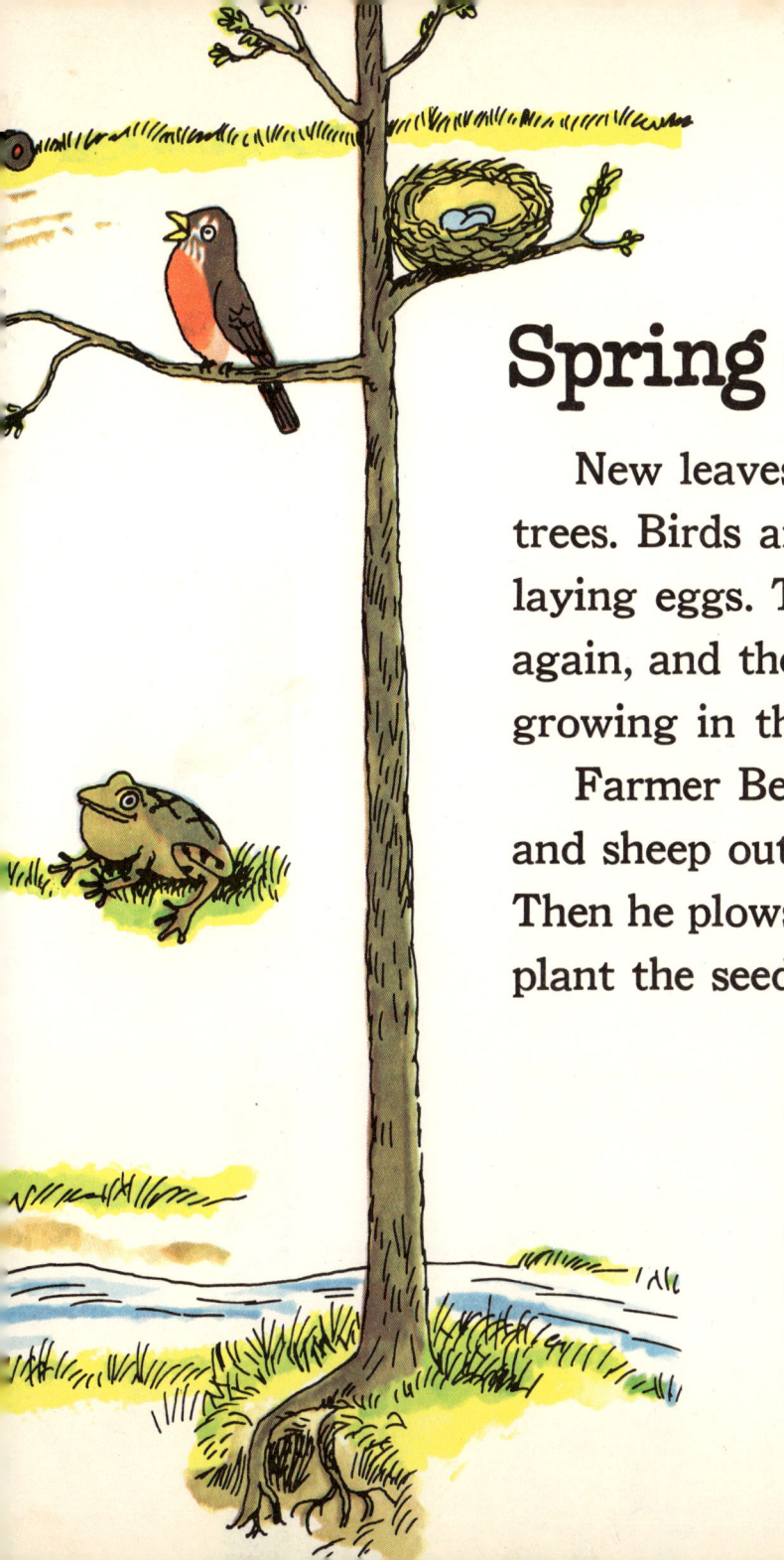

Spring on the Farm

New leaves are growing on the trees. Birds are building nests and laying eggs. The grass is green again, and there are wildflowers growing in the fields. It is spring.

Farmer Bear lets his cows, horses, and sheep out into the pasture. Then he plows his fields. It is time to plant the seeds.

Point to:

lamb	bird's nest
bridge	stream
tree	tractor
robin	new leaves
frog	flowers

Do you see:

pond	dock
log	fence
pasture	cattails
rock	calf
	fishing rod
	Little Fox
	cow
	corn growing in the fields

Summer on the Farm

In summer, the weather is warm. Days are long and nights are short. The grass in the pastures and the leaves on the trees are a beautiful rich green. Crops are growing in the fields.

Little Fox is home from school. He helps on the farm, but he has time to fish or play in the afternoon.

Point to:

gate	basket	pumpkins
apples	squash	roadside stand
wall	Raccoon	fir trees
leaves	corn	apple juice

The weather is cooler. The leaves on the trees change color and begin to fall. It is time to harvest all the crops of summer.

Autumn on the Farm

Raccoon makes fruit juice and jam to sell at the roadside stand along with the squash, pumpkins, corn, and apples from Farmer Bear's fields.

The days are getting shorter. Farmer Bear and his helpers are getting ready for winter.

Winter on the Farm

In winter, when the days are short and the nights are long, it gets very cold. There is frost and the pond freezes over. Farmer Bear brings his animals in to the barn.

Snow falls from the low clouds and all the pastures and fields turn white. Farmer Bear and his friends put on warm clothes. They go out for a sleigh ride. They see little Pig making a snowman. They see Mr. Fox clearing the snow from the road to town.

Do you see:

- scarf
- sleigh
- snowman
- pipe
- snow
- warm hats
- three piglets
- carrot
- snowplow
- Farmer Bear

So the seasons go on the farm. Soon the snow will melt and spring will be back. Then will come summer, then autumn, then winter again.

Through all the seasons, Farmer Bear will care for his land and his animals.

He will get eggs from the chickens, milk from the cow, vegetables from his garden. He has everything he needs on his farm.

He is a happy bear.

THE END